WRITERS REPUBLIC

GEOMETRY OF HINDU TEMPLES

Geometrical Foundations of Hindu Temples

Dr. Salilesh Mukhopadhyay

WRITERS REPUBLIC L.L.C.
515 Summit Ave. Unit R1
Union City, NJ 07087, USA

Website: *www.writersrepublic.com*
Hotline: *1-877-656-6838*
Email: *info@writersrepublic.com*

Ordering Information:
Quantity sales. Special discounts are available on quantity purchases by corporations, associations, and others. For details, contact the publisher at the address above.

Library of Congress Control Number:		2023900369
ISBN-13:	979-8-88810-202-2	[Paperback Edition]
	979-8-88810-203-9	[Hardback Edition]
	979-8-88810-204-6	[Digital Edition]

Rev. date: 01/24/2023

"LEAD ME FROM THE UNREAL TO THE REAL,
FROM DARKNESS TO LIGHT,
FROM DEATH TO IMMORTALITY"

Brhadaranyaka Upanisad

Dedicated to
All those I loved

Contents

INTRODUCTION

Some quarter of a century back when I migrated to America from Australia it was my dream to publish this book. Ultimately at last I'm confident that the book is publishable. In early nineties of the last century, I stayed in Calcutta for a while and was involved with Calcutta Mathematical Society. One of the projects sponsored by the Indian National Science Academy [INSA] was "Excellence of Hindu Mathematics during Indus Valley Civilization". My involvement with the project was limited to the hydro dynamical engineering to regulate the flow of the Sindhu-River to the bath houses of Mohenjo-Daro and Harappa. It's quite challenging.

From the dawn of civilization, it was noticed that the phenomenon started from the banks of a River. Thus, the word HINDU came from the river "SINDHU". The religion started from the banks of the Sindhu River. Just like the European civilization started from the banks of the rivers Tigris and Euphrates.

Although I will talk about Hindu Temples and the Geometry behind the architectures my motivation for this book came from the mathematics behind the Earthquake-Proof Architectures. The science behind these constructions developed in the last century. Specifically in the decades of Sixties and Seventies we noticed the construction of earthquake-proof architectures-based skyscrapers in America and Japan.

While in Iceland I noticed and experienced the three types of earthquakes caused by the volcanic eruptions or the movement of the tectonic plates or Tsunami. India is the land of "Maharajas". The palaces they built has a geometry of its own based on the centuries old "Vastu

Shastra". This is a Sacred geometry which dictates the blueprint of the entrances to the Palace and its dwelling quarters, kitchen, bathrooms etc.

All these dynasties had their own beliefs and accordingly they constructed number of temples for the worshipping of the deities. "GARBHA- GRIHA" is the place where the idol stays in any temple. As there are thousands of temples in India this could be a multivolume compendium of the geometry behind the architectures. Here I will only elaborate the principal hypothesis behind these constructions. "NO ONE DEVOTEE CAN STEP ON THE SHADOW OF THE TEMPLE"! This is a tall demand depending on the height of the temples. The science of measuring shadow started from the study of Solar and Lunar Eclipses. The centuries old three- body problem. Also, the "Height and Distance" in mathematics enables us to measure the actual height. Moreover no one can step on the top of the temple. So, they call that part from Sanskrit "SIKHAR", where they usually put a flag or ornament called "DHAWJA", KONARK in Orissa, INDIA had a huge magnet causing shipwreck.

By the way there are variations on the constructions of these temples. Noticeable distinction is "SHRAVAN-BELLEGOLA" in Mysore. Here you will see the open-air architecture. Just the naked statue of Lord Tirthankara is carved on the side of the mountain. Exceptionally brilliant and marvelous construction of KONARK TEMPLE in Puri, Orissa. Here the dynamical aspect of the Hindu religion is reflected with "RATHA-CHAKRA" THE WHEELS OF THE CHARIOT FOR SUNGOD.

If you notice carefully the flag of India, there are twenty-four spindles in the ASHOKE-CHAKRA to remember twenty-four hours of the day. Now we know similar method of construction was used to build ancient ANGKOR in Cambodia. On summer solstice the ray of the sun comes inside the temple through the door. It should be mentioned here that in Hindu religion we worship the Sun and observe the lunar calendar.

Ancient civilizations always had a stone monument. Like the Mayan had CHECHENITZA in Mexico, Ayers Rock in Australia for the Aboriginals, Mount Rushmore for the American Indians and Stone

Henge for British. India has Taj Mahal. Recently archeologists found a Stone Henge-type architecture in Spain, too.

Long before the birth of Lord Jesus the caves of Ajanta and Ellora were carved and are still standing tall. I personally visited those caves in Aurangabad. What a brilliant idea to carve the temples or caves on a rigid rock which is earthquake proof. During the reign of Emperor Ashoke there were several "STUPA" built for Buddhist religion. Famous one is in Sanchi, Madhya Pradesh, INDIA.

Ajanta caves depicts wall paintings of Buddhist relics and monasteries near Ajanta village in Aurangabad district of the State of Maharashtra. Two and a half thousand years back they carved the mountain with granite cliffs on the inner side of a 70-ft [20-m] ravine in the Wagurna River Valley, approximately 65 miles [105 Kilometers] north-east of Aurangabad.

The cluster of 29 caves of AJANTA was excavated between 1st. century BC and the 7th. Century AD and consists of two types, CAITYAS (Sanctuaries) and VIHARAS (Monasteries). It is the frescotype paintings that are the main attractions of the Ajanta caves. The rich ornamentation of CAITYA pillars is noteworthy. The inherent geometry of carving the mountain with chisel and axe is commendable. According to the New Encyclopedia Britannica Volume 1,"these paintings depict colorful Buddhist legends and divinities with an exuberance and vitality that is unsurpassed in Indian art."

The 29 caves of AJANTA were accidentally rediscovered in the 19th century, while 34 caves of ELLORA remained open to the world, throughout the Middle Ages.

AJANTA caves were carved out from the 200 B.C. to 650 A.D. They were scooped out into the heart of the rock so that the pious Buddhist monk could dwell and pray. AJANTA is world famous for its multi-color fresco pictures and writings in DEVNAGARI. It's dedicated solely to BUDDHISM, HINDUISM and JAINISM. The believers of millions of INDIAN.

ELLORA caves are cut into the sides of basaltic hill. The caves are carved during 350 A.D. to 700 A.D.

THE CAVES AT ELLORA

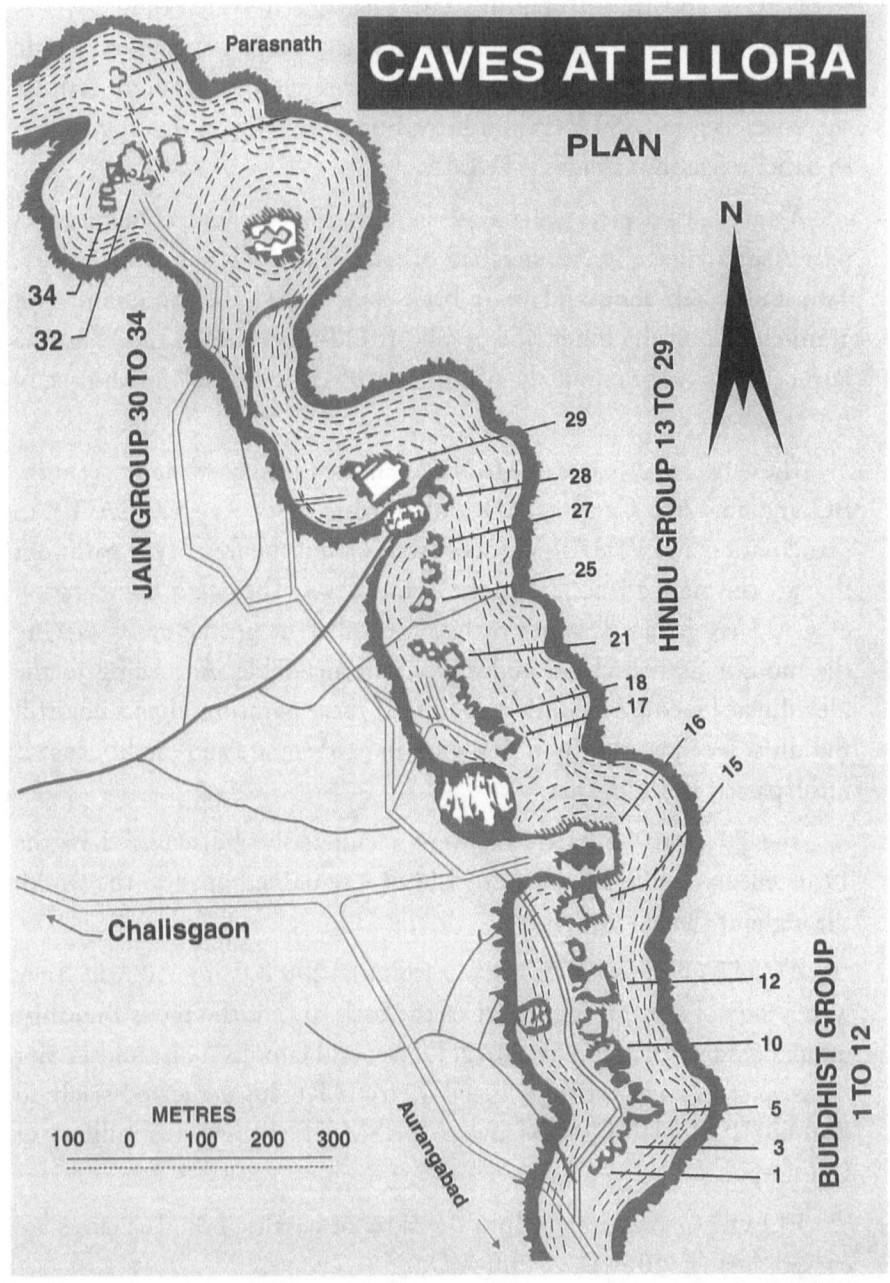

GEOMETRICAL PATTERNS OF AJANTA CAVES

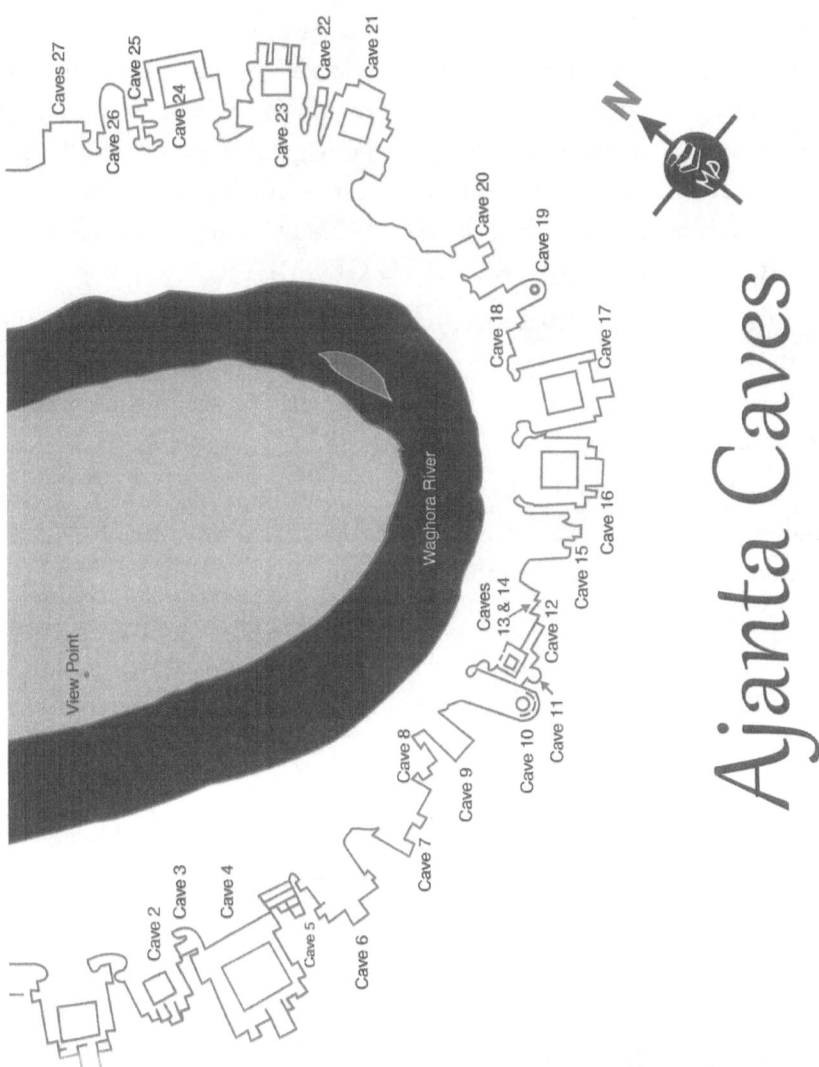

Thanks to UNESCO to declare these caves as World Heritage sites. Now I will talk about the first word in the title of this book-- "GEOMETRY"!

The axiomatic foundation of geometry was led by Euclid around 300 B.C. He probably received his early mathematical education in Athens, Greece from the pupils of Plato, since most of the geometers and mathematicians on whom he depended on were of that school. The only other fact concerning Euclid is that he taught and founded a school at Alexandria in the time of Ptolemy I, who reigned from 306 to 283 B.C. Proclus claims that it was Ptolemy I who asked Euclid if there was no shorter way to geometry than the "ELEMENTS" and received as answer: "THERE IS NO ROYAL ROAD TO GEOMETRY."

The Vastu Geometry is based on Euclidean geometry and the stability of the constructions adhere to some specific shapes and sizes which can be analyzed using fractals, where the dimensions are not integral but fractional.

I will not go into the details of Basic concepts of hour, days, weeks, and months but it was Hindu mathematicians who first found the number EIGHT [8] as the most auspicious.

Accordingly, they divided the day comprising of twenty-four hours into eight "PRAHARS ". The first "PRAHAR" is counted after midnight. Thus 12 mid-night to 3am is the first PRAHAR and so on. Similarly, the Hindu Mathematicians found EIGHT [8] essential parts of the human body.

Thus, the eight parts of the human body consists of two hands [2], Heart,[1], Forehead, [1], two eyes [2], throat [1], and spine or backbone [1]. Alternately, two toes [2], two knees [2], two hands [2], chest [1] and nose [1].

While worshipping the deity one should use the whole body by engaging in these eight parts. There are Eight-fold simple yoga. The first one is Celibacy, [Jom], Rules from the scriptures, [Niyom], comfortable sitting position [Asana], Breathing exercise [Pranayam], Rejection of worldly materials, [Pratyahar], Concept [Dharona], Meditation [Dhyan] and the ultimate enlightenment [Samadhi].

Even they found eight forms of Lord Shiva. These eight forms of Lord Shiva are SARBO [Everywhere], BHAVO [Holds], RUDRA [Excited], UGRA [Annoyed], BHIMA [Powerful], PASHUPATI [Lord of the Animal kingdom], MAHADEVA [The great God] and ISHAN [Lord Shiva].

The Vaishnav sect of Hinduism who believe in Lord Vishnu started chanting the holy script for twenty-four hours meaning EIGHT-PRAHARS.

Even they started constructing Hindu temples which have eight sided roofs in Bengal. Here one can notice the unique style of detached temples of Lord Shiva in a row. These temples are mainly seen in the district of BANKURA and VISHNUPUR of the present-day West Bengal or PASHIMBONGO.

Some of the temples defy all the usual characteristics and are named by Bahai faith as the "LOTUS TEMPLE' in New Delhi. Here you will find no deity as such but the scriptures from different faith and the temple is in the shape of the Lotus flower.

Similarly in Amritsar one will find the famous Golden temple. Here the temple is covered with very thin gold foils and the geometry is extremely intricate. In a sense it has solved the covering problem of mathematical topology by using square plates. Here also there is no deity. Only the Holy book of Sikhism – "The GURU-GRANTH SAHIB."

Interesting observation of the location of the Lord Shiva's famous temples one can draw the line from North of India to South of India and find eleven temples starting from KEDARNATH to RAMESWARAM are all situated along 79-degree latitude.

MEENAKSHI TEMPLE, MADURAI, INDIA

Every country puts its unique signature on stone monuments or
stone carvings! Thus, the Ayer's Rock, the largest monolithic rock in
Alice Springs represents AUSTRALIA, the stone-henge, a prehistoric
megalithic structure on Salisbury Plain, Southeast Wiltshire, and
England represents the UNITED KINGDOM, INDIA has the Taj

Mahal. UNITED States of America has the Mount Rushmore in the Black Hills of Western South Dakota on the side of which are carved gigantic faces of Presidents Washington, Jefferson, Lincoln and Theodore Roosevelt.

My original field of study is Mathematics, Statistics, Mathematical Physics and Computer Science. So, I decided to dig deeper into the Geometry of Hindu Temples. By the way the ancient Vedic geometry has a branch called Vastu Geometry with Vastu Angle etc. In this last book: GEOMETRY OF HINDU TEMPLES you will see the idea and the extensive study on Soil Mechanics by the Hindu Mathematicians to construct temples and caves, stupas with the earth- quake proof architecture.

"SHIV SHAKTI AKSHA REKHA" I can see Somnath & Varanasi at two ends. Shakti Peeths like Kamaksha in Gauhati, Kali Temple in Kalighat, Ambaji in Gujrat. All these temples have unique features in the geometrical analysis. Furthermore, the following eight SHIVA

TEMPLES:

1. KEDARNATH 79.0669 Degrees
2. KALAHASTI 79.7037 Degrees
3. EKAMBARNATHA-KANAHI 79.7036 Degrees
4. THIRUVANAMALAI 79.0747 Degrees
5. THIRUVANAIKAVAL 78.7108 Degrees
6. CHIDAMBARAM NATARAJA 79.6954 Degrees
7. RAMESWARAM 79.3129 Degrees
8. KALESHWARAM 79.9067 Degrees Latitude! By the way it was not simple coincidence!

Mathematicians of India found out the most stable line of latitude on earth through INDIA!!!!

These temples are hundreds of miles apart and it was very difficult to explore the vast expanse of INDIA on those days. Here I will highlight several key points for the underlying geometry of Neoclassical architecture. This book is the preamble of a series of extensive geometrical analysis of the Hindu temples of INDIA and the neighboring regions, which was the greater undivided INDIA. Some of these temples have a huge kitchen like in the Golden Temple of Amritsar and Jagannath temple in Puri. Geometry for the construction of these buildings is simple and it has withstood many centuries.

In this context it should be noted that all these centuries and millennia old temples and caves kept no record of the architect/ architects involved in the construction. The record only shows in history who financed these huge projects.

But one thing for sure just like in Muslim religion the Mosques and Mausoleums like "THE TAJ MAHAL" are built with a typical geometrical measurement the construction reflects the period of the architecture. In the construction of European palaces and churches we have GOTHIC ARCHITECTURE, BAROQUE ARCHITECTURE etc. In India it's difficult to identify the UNIQUENESS of temples. For example, the KONARAK Temple is built with a dynamic structure on a Chariot. THIS IS THE SECOND ARCHITECTURE WITH AN EPOCH-MAKING CONCEPT OF WORSHIPPING THE SUN AND THE SOLAR RAYS. THE MARTAND TEMPLE IS THE FIRST. Coordinates: 330 44'44" N 75013'13" E. One will never find this type of ideas elsewhere.

According to Kalhana, the MARTAND Temple was commissioned by Lalitaditya Multipedal in the Eighth Century AD. This Hindu temple is located near the city of ANANTANAG in the Kashmir Valley of Jammu and Kashmir (UNION TERRITORY), India. The Sanskritlanguage synonym of SURYA (Sun) is MARTAND. This temple was destroyed by Sikandar Shah Miri (1389-1413) in a zeal to Islamize the society under the advice of SUFI preacher MIR MUHAMMAD HAMADANI.

The MARTAND TEMPLE was built on top of a plateau from where one can view whole of the KASHMIR VALLEY. From the ruins and related archaeological findings, it can be said it was an excellent specimen of KASHMIRI ARCHITECTURE, which had blended the GANDHARAN, GUPTA and CHINESE forms of architecture.

The temple has a colonnaded courtyard, with its primary shrine in its center and surrounded by 84 smaller shrines, stretching to be 220 feet long and 142 feet broad total and incorporating a smaller temple that was previously built.

The temple turns out to be the largest example of PERISTYLE in KASHMIR and is complex due to its various chambers that are proportional in size and aligned with the overall perimeter of the temple. In accordance with Hindu temple architecture, the primary entrance to the temple is situated in the western side of the quadrangle and is the same width as the temple itself, creating grandeur.

The entrance is highly reflective of the temple due to its elaborate decoration and allusion to the deities worshiped inside. The primary shrine is located in a centralized structure (the temple proper) that is thought to have had a PYRAMIDAL top—a common feature of the temples in KASHMIR. Various wall carvings in the antechamber of the temple proper depict other gods, such as VISHNU, and river Goddess, such as GANGA and YAMUNA, in addition to the SUN-GOD SURYA.

The KHAJURAHO GROUP OF TEMPLES have coordinates: 24051'16" N 79055'17" E.

These group of HINDU and JAIN temples are located in CHHATARPUR District, Madhya Pradesh, India, about 175 kilometers southeast of JHANSI.

THEY ARE A UNESCO WORLD HERITAGE SITE.

The temples are famous for their NAGARA-STYLE architectural symbolism and a few erotic sculptures.

Most KHAJURAHO temples were built between 885 A.D. and 1050 A.D. by the CHANDELA DYNASTY.

Historical records note that the KHAJURAHO temple site had 85 temples by the 12th. Century, spread over 20 square kilometers. Of these, only 25 temples have survived, spread over six square kilometers.

Of the surviving temples, the KANDARIYA MAHADEVA TEMPLE is decorated with a profusion of sculptures with intricate details, symbolism and expressiveness of ancient INDIAN ART.

The Khajuraho group of temples belong to Vaishnavism sect of Hinduism, SAIVISM sect of Hinduism and Jainism—nearly a third each. The temple site is within VINDHYA mountain range in Central India. Legends have it that Lord SHIVA and other Gods enjoyed visiting the dramatic hill formation in Kalinjar area. The center of this region is KHAJURAHO, set midst local hills and rivers. The temple complex reflects the ancient Hindu tradition of building temples where Gods love to play.

Architecture of the Khajuraho temples -
(exemple of Kandariya Mahadeva temple)

Amalaka

Kalasha

Sikhara
(tower)

Urushringa
(Subsidiary sikhara)

Antarala
(Vestibule)

Garba griha (Shrine,
inside the sikhara)

Maha mandapa
(Great hall)

Mandapa
(Hall)

Pradakshina
(circumambulation)

Ardha mandapa
(Entrance porch)

Jagati
(platform)

East

Adhisthana
(base platform)

Transepts

GEOMETRY OF EARTHQUAKE-PROOF ARCHITECTURES

Temples are the landmarks of the Hindu faith and is constructed in a way which is unique for the period. There was no cement on those days. They used stones, bricks, sand, SURKI and Calcium dust to have a rigid structure. Mostly the GARVA-GRIHA are designed as a square room, but the GOPURAM is circular but conical. So, there are concentric circles of different radii, the largest is at the bottom and the smallest is at the top.

Hindus believe that before entering the temple one should take a bath and cleanse oneself externally with body soap and shampoo. While in the temple they will cleanse their interior mind, body and soul. Ponds or lakes are mostly attached to the temple. Some deity has a bathing ceremony, like The JAGANNATH DEV. Here in Puri, Orissa the Lord Jagannath is bathed properly and then put the new dress to visit "GUNDICHA BARI" on a specially decorated chariot drawn by humans.

This entire process is a religious ritual every year and is called "RATHAJATRA."

Rigid mountains are the best examples of earthquake-proof architectures. Here the geometry of carving a cave is simple. But imagine the difficulty in getting those fragments [Tons and Tons of rubbish] out of the cave was not an easy task. Especially there were no power tools.

Here I would like to mention similar carvings of Christian Churches were found in Sudan. Styles of carvings are starkly different, but the structures were earthquake-proof. These Churches were covered in 60 Minutes the CBS program.

Already in the Introduction I have mentioned about the Caves of Ajanta in Aurangabad, INDIA. Not far from there a cluster of 39 caves were excavated in ELLORA. Ellora caves were carved in the GUPTA PERIOD [6th-8th century AD] near the village of ELLORA, Aurangabad district, Maharashtra State, western INDIA. These caves are 18 miles [29 Kilometers] from the town of Aurangabad. These caves have Hindu deities like SHIVA, Jain deities and different positions of Lord Buddha.

The monolithic KAILASHNATHA TEMPLE of Lord SHIVA [who resides in the mountain named Kailash, a peak in the Himalayas] is 165 ft. (50 m) long and 96 ft. (30 m) high and is world famous stone curved monument. This was built in 8th century during the reign of RASTRAKUTAS.

The other caves have chisel-curved elephants and different animals. Remember every HINDU Gods has a carrier which is an animal. Like

Ma Durga always rides a lion, Ma Saraswathi always rides a duck, Ma Lakshmi an owl and so on.

Sometimes in Hindu temples one will observe the "NABA-GRAHA MANDIR" [The temple for nine planetary deities] inside the main temple. These deities are **SUN ["Surya" on a horse-drawn chariot], MOON ["Chandra" on a pedestal], MARS ["Angaraka"- Mangal], MERCURY ["Buddha"], JUPITER [Brihaspati, Guru], VENUS [Sukra], SATURN [Sani], ECLIPSES [Rahu emerging from clouds], COMETS & METEORS [Ketu on a lion]. The last two, RAHU and KETU are shadow planets representing the ascending and descending of lunar nodes.**

Neptune, Uranus and Pluto were not discovered or recognized to have an influence on humans.

The geometry of these planetary deities is a **SQUARE with three deities on each side.**

There are TEN **"DIKPALAS"** [Guardians of direction] in Hindu religion:

EAST: INDRA on the three-headed elephant AIRAVATA, God of the sky.

SOUTHEAST: AGNI, on a ram but in Khmer art [ANGKOR WAT] a rhinoceros, God of Fire.

SOUTH: YAMA, God of judgment and hell SOUTHWEST: NIRRITI usually on a human being, Goddess of death and corruption.

WEST: VARUNA on a water monster [MAKARA, but sometimes a NAGA, God of the Ocean.

NORTHWEST: VAYU, on an antelope, God of the Wind.

NORTH: KUBERA on the magical chariot of PUSHPAKA, God of Wealth.

NORTHEAST: ISHANA, an aspect of SHIVA on a bull, God of Air and Wind [In this role a DIKPAL not as a destroyer].

URDHA: ZENITH – Lord Brahma!

ADHO: NADIR – Lord Vishnu!

Here the geometry is a solid sphere or a solid cube, meaning the entire universe.

GEOMETRY OF KHMER ARCHITECTURE

The characteristic of Khmer architecture is the symmetry. The temples were not designed as a place for meeting rather than the dwelling house of the deities. During festivals they used to get together in the temple arena which was sprawling. The SIKHARAS were extremely symmetrical. One of the weapons of Lord Shiva is the metal trident [TRI (meaning three) and SHUL (meaning Piercing ability)]. The top of the temple had three tall structures mimicking the TRIDENT of Lord Shiva.

Geometry of construction for temples suggest that a cluster of small temples is more stable during earthquake than a single tall temple. Interestingly enough base of the temple should be the largest frame upon which the GOPURAM rests. The shape of the GOPURAM should be cylindrical as shown in the following figures.

The temples in those days were built by simple scaffoldings and there was no crane. So the builders were finishing the ground level work and then climb on the top of that roof and start a fresh construction of the next phase.

It should be noted here that most of the temples have a single "GARVA-GRIHA". During "Hoysal" dynasty of Karnataka, South India a special design of architecture was introduced in the "LAKSHMINARASINGHA" temple at "NAGEHOLLI". This temple has THREE "GARVA-GRIHA."

In the western GARVA-GRIHA has the deity "CHENNAKESHAB", in the south GARVAGRIHA the deity of "BENUMADHAV" and in the North GARVAGRIHA the deity of "LAKSHMI NARASINGHA."

The three doors of the three GARVAGRIHAS open at a single "NAVARANG MONDAP." This temple also has other deities like Ma Durga, Ma Saraswathi.

One should notice the flexibility of Temple construction. Although it's a rule that one "GARVA- GRIHA" per "GOPURAM" here we see THREE "GARVA-GRIHA" for one GOPURAM. Analytical minds can explain it very clearly. In HINDU religion everything is related to what a HINDU does.

Just like in human pregnancy people expect a single child. But in human birthing there are twins, triplets, quadruplets and so on. Thus the "GARVA-GRIHA" in each temple is as auspices as a woman's womb. Concept of multiple pregnancy explains the geometrical design of multiple "GARVA-GRIHA" – all opening in the same "MANDAP."

Here I cannot resist the citation of "SRI VARADHARAJA PERUMAL" temple. This was completed in 3rd. BC.

Here in the insertion one can notice the Chain link hewn made from curving stones. Not an easy task! Especially the application of "TORUS" or "DOUGHNUT" in topological language as we understand now was not developed.

First thing first. One should ask why on earth Hindus designed this ornament on the temple. Remember Chain-Pully method of lifting heavy materials was invented by Archimedes a few years back and the idea was extensively used to bring water from the wells in India. It was believed that in earthquake if the structure goes down and the stone-curved chains are intact the structure can be salvaged very easily. Moreover, chains can be used to arrest the movement. So, if there are chains made of stone are placed on the temple the deity will never move and temple will be functioning for years to come.

Different dynasties had left their unique architectural signatures on the construction of temples in India. Let us consider next the famous 'GANGAIKONDA CHOLAPURAM" temple in "Gangaikonda Cholapuram", Tamil Nadu, India. This is a grand structure which reflects our glorious past. This was built over a thousand year's back [1,000] during the reign of "CHOLAS." This gigantic 180- feet-tall "GANGAIKONDA CHOLEESWARAM MANDIR" [Temple] shows the architectural prowess of the "CHOLAS" and needs elaborate geometric analysis.

In Hindu religion one can also worship the birds. Thus "GARUR" the largest bird is the carrier of Lord VISNU. "PAKSHEE TEERTHAM" in Tamil Nādu, INDIA has the centuries old temple of Lord SHIVA and Egyptian vultures used to visit the place regularly till 1998. On the top of the mountain a few steps lead to the ledge where the bird-feeding takes place. I've noticed similar type of bird-feeding in the KRUGER National Park in South Africa.

According to the Encyclopedia Britannica, on the eve of the Muslim occupation, the Hindu religion was by no means sterile in northern India, but its vitality was centered in the southern. Dravidian-speaking areas rather than in the north. Blatant discrimination based on the complex hierarchy of the castes, strictly forbidden intermarry and interdine, even stepping on the shadow of a Brahmin were accepted without any question. Child marriage, polygamy, and widow burning were widespread.

The Gupta Empire was an ancient Indian empire which existed from the early 4th.Century CE to late 6th Century CE. At its Zenith, from approximately 319 to 467 CE, it covered much of the Indian subcontinent. THIS PERIOD IS CONSIDERED AS THE **"GOLDEN AGE"** OF INDIA BY HISTORIANS.

From the **"GUPTA PERIOD"** onward temples tended to become larger and more prominent in Hinduism, and their architecture / Geometry developed in distinctive regional styles. In northern India the finest remaining Hindu temples are to be in Orissa, as well as at

KHAJURAHO, in northern Madhya Pradesh. The finest example of Orissa temple architecture is the "LINGARAJA TEMPLE" of BHUBANESWAR, Orissa built in c.1000.

The largest temple and a unique architecture on a CHARIOT [To bring the dynamics of propagation of Sunrays to the Earth] of the region, however, is the famous "BLACK PAGODA, the "SUN TEMPLE" (SURYA DEULA) of KONARAK, built in the mid-13th century. Its 200 ft. (60 meters) tower has long since collapsed, and only the assembly hall remains.

The most important KHAJURAHO TEMPLES were built during the 11th century. Individual architectural styles also arose in GUJARAT and RAJASTHAN, but I'm not elaborating the famous "SOMNATH TEMPLE" and the others.

By the end of the first millennium AD the South Indian style had reached its apogee in the GREAT "RAJARAJESVARA TEMPLE" of THANJAVUR [TANJORE].

The designing of Hindu temples, like that of religious images, was codified in the "SILPA- SASTRAS" (CRAFT TEXTBOOKS) and every aspect of the design was believed to be symbolic of some feature of the cosmos. The idea of microcosmic symbolism is very strong in Hinduism and comes from the Vedic times. It is in this context that the erotic sculptures carved on the outer walls of the towers of some medieval temples, notably at KAHAJURAHO and KONARAK, may perhaps be interpreted.

SIMPLIFIED SCHEME OF A HINDU TEMPLE:

GOPURAM: TOWER-GATE Here 2 but more often 4, one on each side. The main deity or Shrine is situated in the "GARBHA GRIHA."

Along the axis West-East the "JAGA MOHAN", MANDAPA: Meeting Hall

"NAT-MANDIR": Dance Pavilion

"BHOG-MANDIR": Offerings Hall

The Hindus built all these temples to worship their deities in the same fashion they lived their lives. The rituals are almost identical but more elaborate and divine. In the temple the God was worshipped by the rites of PUJA (reverencing a sacred being or object) on the analogy of serving a great king. In the important temples, a large staff of trained officiates waited on the God.

He was awakened in the morning along with his goddess, washed, clothed and fed, placed in his shrine to give audience to his subjects, praised and entertained throughout the day, ceremoniously fed, undressed, and put to bed at night.

Worship consists of hymn singing, burning of lamps and perfumed candles/ sandalwood sticks, waving of special oil burning candles before the divine image (ARATI) and similar acts of homage.

The God's dancing girls [DEVADASIS] would perform before him at regular intervals, watched by the officiates and lay worshippers, who were his courtiers. The DEVADASI culture of Hindu temples has long been abolished.

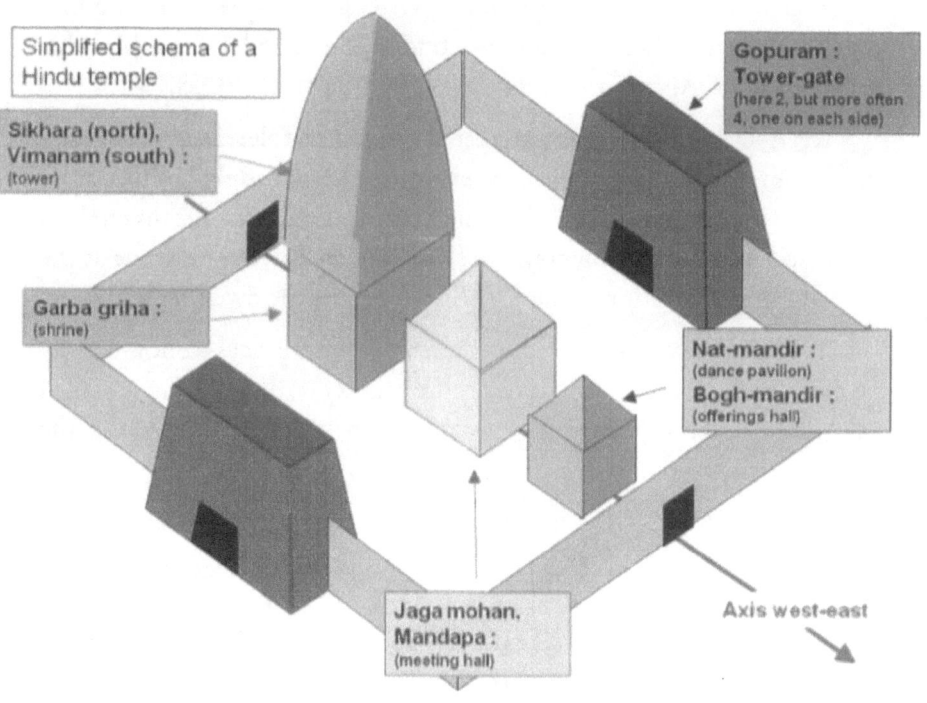

Simplified schema of a Hindu temple

Sikhara (north), Vimanam (south) : (tower)

Garba griha : (shrine)

Jaga mohan, Mandapa : (meeting hall)

Gopuram : Tower-gate (here 2, but more often 4, one on each side)

Nat-mandir : (dance pavilion) Bogh-mandir : (offerings hall)

Axis west-east

THE VASTU GEOMETRY

There are essentially two types of VASTU GEOMETRY. The first one is followed in INDIA and the second type of VASTU GEOMETRY are used in JAPAN to construct the BUDHIST TEMPLES.

KHAJURAHO temples, almost all Hindu temples designs, follow a grid geometrical design called **"vastu-purusha-mandala"**. This design plan has three important components – "MANDALA"— means circle as discussed in Hindu Geometry. "PURUSHA" is universal essence at the core of Hindu tradition, while "VASTU" means the dwelling structure.

The design lays out a Hindu temple in a symmetrical, concentrically layered, self-repeating structure around the core of the temple called "GARBHAGRIHA' where the abstract principle PURUSHA and the primary deity of the temple dwell.

The SHIKHARA, or spire of the temple rises above the "GARBHAGRIHA". THIS SYMMETRY AND STRUCTURE IN DESIGN IS DERIVED FROM CENTRAL BELIEFS, MYTHS, CARDINALITY, AND MATHEMATICAL PRINCIPLES, Vide: Stella Kramrische [1976] The Hindu Temple Volume 1, ISBN 81-208-0223-3.

THE CIRCLE OF MANDALA CIRCUMSCRIBE THE SQUARE. THE SQUARE IS CONSIDERED DIVINE FOR ITS PERFECTION AND AS A SYMBOLIC PRODUCT OF KNOWLEDGE AND HUMAN THOUGHT, WHILE CIURCLE IS CONSIDERED EARTHLY, HUMAN AND OBSERVED IN EVERYDAY LIFE [THE MOON, THE SUN, THE HORIZON, THE WATER DROP, EVEN RAINBOW]. THIS IS VERY MUCH LIKE THE CONCEPT OF STRONGER NEIGHBORHOODS IN TOPOLOGY.

THE SQUARE IS DIVIDED INTO 64 SUB-SQUARES CALLED "PADAS".

THIS MAGIC NUMBER 64 = 8 X 8 IS THE SOURCE OF THE THEORY OF CHESS. MOREOVER IN "KAMA-SUTRA" YOU WILL FIND 64 VARIETIES OF EROTIC POSITIONS [OR KALA]. ULTIMATELY, WE SEE A COMMON REPEATATIVE PATTERN. SIMILAR IS THE ANALYSIS OF 81 = 9 X 9 GRID. SMALLER TEMPLES USE 9, 16, 36 OR 49 GRID MANDALA PLAN.

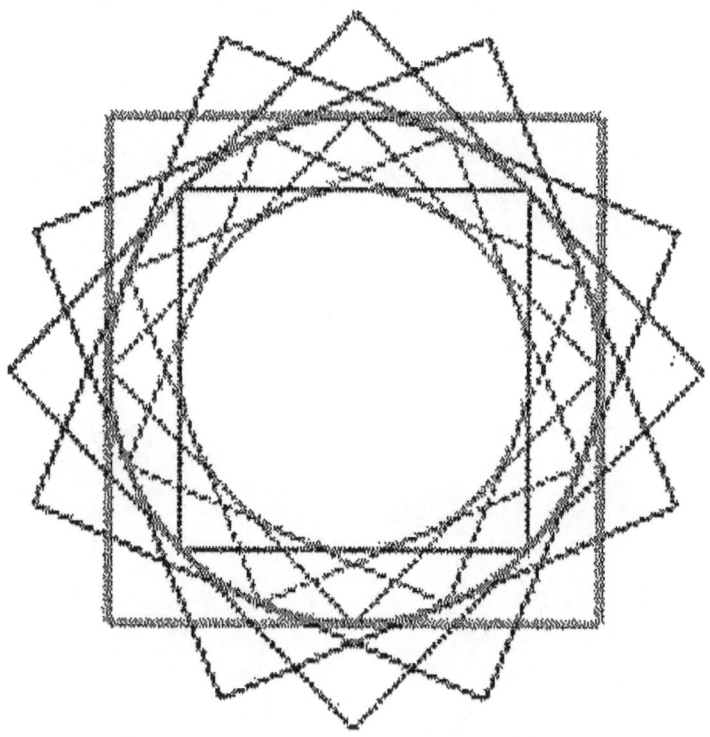

HINDU GEOMETRY Versus EUCLIDEAN GEOMETRY

We are not quite sure how the Euclidean geometry was brought to India, but we are confident that "SULV SHASTRA" or "Mathematics of lines [REKHA]" were developed as "HINDU GEOMETRY." Sometimes in Bengali we called this branch of Mathematics as "JYAMITI" MEANING MEASUREMENT OF EARTH! This term "JYAMITI" is very interesting because it is very alike the Greek term "Geometry", not only phonetically but also in significance, and at the same time it is not a Hinduised Greek word. The compound word "JYAMITI" in Sanskrit is derived from "JYA", meaning "EARTH" and "MITI", meaning "MEASURE". Hence its literal significance is "Earth Measurement."

According to Datta, B, Singh, A.N. and Shukla, K.S [1979] in the Vedic ages there were three notable schools of **Baudhayana, Apastamba and Katyayana.** Although the results obtained were the same but the individual school had its own method of derivation. The Vedic Geometry, as found in the manuals of the "SULBA", was not wholly intuitional. In fact, we find a kind of proof in case of certain propositions of the "SULBA."

For example, how to find the area of a trapezium, has been demonstrated by "Apastambha" in the course of mensuration of the "MAHAVEDI", which is of the shape of an isosceles trapezium whose altitude, face and base are respectively 48, 24 and 30 "PADAS"[or "PRAKRAMAS"].

Here the most important thing to be noted is there was no unit of measure of modern days like the FPS [Foot, Pound, Second] or MKS [Meter, Kilogram, Second]. Apastamba says:

"The "MAHAVEDI" measures in area one thousand less twentyeight Squares PADAS. Draw a straight line from the south-eastern corner of the VEDI to a point 12 PADAS toward the south- western corner. Place the portion thus cut off on the other (i.e., the northern) side of the VEDIMAHAVEDI) will then be a rectangle. After that construction the area will be apparent."

Vide APASTAMBA SULBA [Vol. 7]

After the so-called Pythagoras Theorem, BAUDHAYANA observes the truth of it.

H2 + (12)2 = (30)2. Or H =27 PADAS. Now 27 X 36 = 972 which is 28 Squares less than thousand Squares PADAS.

After describing the constructions necessary in a proposition, the early HINDU Geometers are found to have remarked "SAMADHIH" [or "This is the construction"].

The expression "QUOD ERAT FACIENDUM" [or "What was required to do"] occurring at the end of a proposition of EUCLID's ELEMENTS.

LATER HINDU GEOMETRY

In later Hindu Geometry, the Hindus undoubtedly showed considerable proficiency in mensuration of certain rectilinear figures and indeed they obtained some remarkable results, e.g. a new proof of Pythagorean Theorem, formulae for the area and diagonals of an inscribed convex quadrilateral and rational solution of triangles and cyclic quadrilaterals. There was no definition, no postulates, no axioms, no proofs of theorem, simply nota formal foundation of geometry. It is noteworthy that the later Hindus included geometry to their treaties of arithmetic [PATIGANITA]{"PATI" meaning a special type of Mat and "GANITA" meaning MATHEMATICS}. Because there were no chair tables Hindus used sit on the floor with "PATI" and used to do Mathematics. More specifically in the section of "KSETRA" [Plane

28

figures], "KHATA" {"Excavations"}, "CITI" ["Piles of Bricks"], "RASI" ["Mounds of Grain"] and "KRAKACIKA" ["Saw"]. The last four topics are pertaining to solid figures.

EUCLID'S ELEMENTS IN INDIA

From the fundamental development of Hindu Geometry, it is evident that it has no resemblance to Euclid's ELEMENTS. The earliest mention of Euclid's Elements was found in the writings of "AL-BIRUNI" the eminent Persian mathematician and traveler in 973.

It was during the Muhammadans rule in India towards the close of twelfth century of the Christian era Arabic and Persian works on mathematics began to be brought into India.

King Firuz Shah Bahmani [1397 – 1422] used to hear on three days in a week, lectures on Botany, Geometry and Logic Vide Law, N.N. [1916].

A son of Daud Shah was very fond of "TAHRIR-U-UQLIDAS" [Euclid's ELEMENTS] and used to teach it regularly to his students. The Emperor AKBAR [1575] included it into the course of study for the school for boys. By the way there were no formal education system for the girls on those days. In his "AIN-I-AKBARI, Abul Fazl [1590] has referred to a few propositions of the ELEMENTS in a way which shows his thorough acquaintance with the work. The work, however, remained confined among the Muslim schools in India.

There is no evidence to support the influence of Euclid's work among the HINDUS before the middle of the Seventeenth Century. In 1658 A.D. KAMALAKARA, the court-astronomer of the Emperor JAHANGIR of Delhi, wrote a voluminous treaty on Astronomy entitled "SIDDHANTA-TATTVAVIVEKA". Certain passages in this work appears to be adapted from Euclid's ELEMENTS.

The first complete translation of the work in Sanskrit was made in 1718 A.D. under the title "REKHAGANITA" ["MATHEMATICS OF LINES"] by SAMRATA JAGGANNATHA, under the patronage of King JAYA SIMHA of JAIPUR, INDIA.

Another Sanskrit version is known as the famous "SIDDHANTACUDAMANI". The author of this book is still unknown.

HINDU NAMES FOR GEOMETRY

The Hindu name for the science of geometry has varied from time to time as discussed in detail by Datta, B [1930]. The earliest name was "SULBA". In the early canonical works of the JAINAS [500 – 300 B.C] the more common name for geometry is found to be "RAJJU". The Sanskrit words "SULBA" and "RAJJU" have the identical significance, which is ordinarily "A ROPE" "A CORD." The word "SULBA" [or "SULVA"] is derived from the root "SULB" [or "SULV] meaning "TOMEASURE" and hence its etymological significance is "MEASURING" or "Science of Measurement."

Mention of a linear measure, called "RAJJU" is found in the "APASTAMBA-SULBA" "MANAVA- SULBA" and "ARTHASASTRA OF KAUTILYA" and later in the "SILPA-SASTRA" Precisely in ancient India there were three kinds of measures like LINEAR, SUPERFICIAL and VOLUMINAL – having the same epithet "RAJJU."

In the JAINA canonical works they are sometimes distinguished as "SUCI-RAJJU" [Needle-like or Linear RAJJU], PRATARA-RAJJU [Superficial RAJJU] and "GHANA-RAJJU" [Cubic -RAJJU]. In the "ARTHASASTRA of KAUTILYA the superficial unit of RAJJU is called "PARIDESA" and the cubical unit "NIVARTANA." The use of the word "RAJJU" means a measuring tape for linear measurements.

POST-VEDIC GEOMETRY

The Hindu geometry which started in a big bang way not only by going much in advance of the ancient Egyptian or Chinese geometry but also by anticipating some notable discoveries of the posterior Greek geometry, did not make much progress in post-Vedic period. In this period there was renaissance of Hindu Mathematics vide Datta [1929]. Compared to the challenges of Arithmetic and Algebra, the branch of Geometry seems to have received little impetus for further development. It will not be true to assume that the study of Geometry was completely neglected by the Hindus of the early renaissance period. On the other hand, it is found to have become more popular and was regarded as a part of general education of the people.

In an early JAINA canonical work, composed CIRCA 300 B.C. we find the remark, *"GEOMETRY IS THE LOTUS IN MATHEMATICS,… AND THE RES T IS INFERIOR" Vide B.Datta et al [1979].*

The greatest achievement of this period of Hindu Geometry are the discovery of the Ellipse, [The Conic Sections], Elliptic Cylinder, the value of π = 10. The mention of the Elliptic Cylinder, called GHANAPARIMANDALA or Solid Ellipse in contradistinction to PRATARAPARIMANDALA [Plane Ellipse], occurs in the "BHAGABATI-SUTRA."

In later years, Geometry was called by the HINDUS *"KSETRA-GA NITA"* ["Mathematics of the KSETRA (Plane Figure)"]. This term appears in "GANITA-SARA-SAMAGRAHA" [COLLECTION OF COREMATHEMATICS] of MAHAVIRA [850]. In the mathematical treaties of BRAHMAGUPTA [628], SRIDHARA [900] and BHASKARA II [1150], the section devoted to plane figures is called *"KS ETRA -VYA VA D HA RA"* ("Treatment of Plane Figures").

The epithet "KSETRA-GANITA" occurs as early as the works of SIDDHASENA GANI (550). The term "KSETRA" has a duel meaning in this work to include both areas and volumes. In the same significance it appears in JAINA cosmological works called "KSETRA-SAMASYA". In later publications the two branches of Geometry were treated separately, and "KSETRA-GANITA" meant "The Geometry of Plane Figures."

JAGANNATHA [1718] called his version of Euclid's ELEMENTS the "REKHA-GANITA" [Mathematics of Lines]. BAPUDEVA SASTRI preferred the name "KSETRA-MITI [MEASUREMENTS OF SURFACE AREAS AND VOLUMES"].

Although all along I was referring to "REKHA" by "SARALREKHA" or "RIJUREKHA" meaning "STRAIGHT LINE". But in HINDU GEOMETRY lot of works was done on "BAKRAREK" meaning "CURVED LINES."

In HINDU GEOMETRY, we find two different systems of nomenclature for the rectilinear geometrical figures, Vide DATTA [1930]. In one system the naming is according to the number of sides of the figures and the names are formed by juxtaposition of the number of names with the word "BHUJA" meaning "ARM" [or SIDES]. For example, "TRIBHUJA" [Trilateral or simply Triangle], "CATURBHUJA" ["QUADRILATERAL"] and so on.

In the other system the naming is based on the number of angles and corners in the figures, and the names are compounds of number names with the word "KARNA" or "KONA." The Sanskrit word "KARNA" means the ear.

Applied to geometrical figures, it implies the "ANGLE." "KARNA" also connotes the "DIAGONAL" or the hypotenuse of a right-angled triangle. The word "KARNA" in Sanskrit degenerated into "KONA" in the "PRAKRITA" languages.

Furthermore, triangles are classified according to the sides: "SAMABAHU-TRIBHUJA" [Equilateral Triangle], "DVISAMATRIBHUJA" or "SAMADVIBAHU TRIBHUJA" [Isosceles Triangle] and "VISAMA-TRIBHUJA" [Scalene Triangle].

Similarly, "Circle" is called "VRITTA" and the "Circumference" as "PARIDHI". In the early text "MONDALA" ["Round"] was used for circle. Typical propositions of Hindu Geometry. The construction of "VEDIS" [Alters] dealt with the geometry of the square, Rectangle, the rhombus, the trapezium, the rectangle, the triangle and the circle.

Geometrical Constructions:

To divide a line into any number of equal parts.

To divide a circle into any number of equal areas by drawing chords.

To divide a triangle into several equal and similar areas.

To construct a square equal to the sum of two different squares. To construct a square equivalent to two given triangles.

For details, please consult Datta [1932].

THEOREMS:

The following theorems are either proved analytically or the results are implied in the methods of construction of the altars of different shapes and sizes in different HINDU temples:

Theorem 1

The diagonals of a rectangle bisect each other. They divide the rectangle into four parts, two and two (Vertically opposite) of which are equal in all respects.

Theorem 2

The diagonals of a rhombus bisect each other at right angles.

Theorem 3

An isosceles triangle is divided into two equal halves by the straight line joining the vertex to the middle point of the base. [This line is called the "Median" or "MADHYAMA"].

Theorem 4

The area of a square formed by joining the middle points of the sides of a square is half that of the original one.

Theorem 5

A quadrilateral formed by the lines joining the middle point of the sides of a rectangle is a rhombus whose area is half that of the rectangle.

Theorem 6

A parallelogram and rectangle on the same base and within the same parallels have the same area.

Theorem 7 [PYTHAGORUS THEOREM]

The square on the hypotenuse of a right-angled triangle is equal to the sum of the squares on the other two sides.

Theorem 8 [CONVERSE THEOREM]

If the sum of the squares on two sides of a triangle be equal to the square on the third side, then the triangle is right-angled.

THE BAUDHYANA THEOREM

"The diagonal of a rectangle produces both areas which its length and breadth produce separately."

APASTAMBA and KATYAYANA states the above theorem in almost identical terms. The theorem is now universally associated with the name of the famous Greek geometer PYTHAGORAS (540 B.C.).

The Chinese new the numerical relation for the case 32 + 42 = 52, probably in the time of CHOU KONG [d. 1105 B.C.] Similar integer solution of Fermat's Last Theorem are 82 + 62 = 102 and 162 + 122 =202 etc. As for the HINDUS, one solution of Fermat's Last Theorem 362 + 152 = 392 occurs TAITTIRYA SAMHITA [Before 2000 B.C.]. Here it is to be emphasized that this instance is different from that known to other early nations.

CIRCLE

In early Hindu geometry, the circle was termed "MANDAL" [Round] "PARI-MANDALA" [Round on all sides]; the CIRCUMFERENCE, "PARINAHA" [Surrounding boundary line], the DIAMETER, "VISKAMBHA" or "VYASA" [breadth] and the CENTRE, "MADHYA" [Middle]. The last term was also used to denote the middlemost point of a square, rectangle or line. In PRAKRITA works of the fourth century before the Christian era, the term "PARIMANDALA" is used to denote ELLIPSE.

In later HINDU Geometry, the term for the circle is "VRITTA" and for the center is "KENDRA." The "RADIUS" is called "VYASARDHA" or "VISKAMBHARDHA" [SEMIDIAMETER]. These terms occur as early as the works of UMASVATI [c.150]. SURFACE AND AREA.

In the early INDU geometry, a plane surface bounded by a figure was called by the term "KSETRA" and its area by "BHUMI."

The "SULBA-SUTRAS", which form a part of Vedic literature of the Hindus, deal with the construction of fire alters [HOME-VEDI] for sacrificial purposes. At present we know of SEVEN SULBA-SUTRAS, although it is quite likely that many more such results existed in ancient times.

According to European Scholars, the "SUTRAS" were composed during the period 800 to 500 B.C, but they are actually much older.

The geometry of "VEDIS" [Alters] in these Sutras are very complicated involving Square, rectangle, rhombus, trapezium, triangle and Circle.

The geometrical propositions involved in the constructions are the following:

CONSTRUCTIONS

1. To divide a straight line into any number of equal parts.

2. To divide a circle into any number of equal areas by drawing diameters.

3. To divide a triangle into several equal and similar areas.

4. To draw a straight line at right angles to a given line.

5. To draw a straight line at right angles to a given straight line from a given point on it.

6. To construct a square on a given side.

7. To construct a rectangle of given sides.

8. To construct an isosceles trapezium of given altitude, face and base.

9. To construct a parallelogram having given sides at a given inclination.

10. To construct a square equal to the sum of two different squares.

11. To construct a square equivalent to two given triangles.

12. To construct a square equivalent to two given pentagons.

13. To construct a square equal to a given rectangle.

14. To construct a rectangle having a given side and equivalent to a given square.

15. To construct an isosceles trapezium having a given face and equivalent to a square or rectangle.

16. To construct a triangle equivalent to a given square.

17. To construct a square equivalent to a given isosceles triangle.

18. To construct a rhombus equivalent to a given square or rectangle.

19. To construct a square equivalent to a given rhombus.

THEOREMS

One important feature of the Hindu Geometry is it is mainly developed for the construction of Vedic ALTARS, Temples, Stupas, Platforms for starting fire for worshipping and animal sacrificial ceremonies.

The following theorems are either explicitly stated or the results are implied in different geometric configurations under construction:

<u>Theorem 1</u>

The diagonals of a rectangle bisect each other. They divide the rectangle into four parts of which the vertically opposite parts are EQUAL in all respects.

<u>Theorem 2</u>

The diagonals of a rhombus bisect each other at right angles.

<u>Theorem 3</u>

An isosceles triangle is divided into two equal halves by the line joining the vertex to the middle point of the base.

<u>Theorem 4</u>

The area of a square formed by joining the middle points of the sides of a square is half that of the original one.

<u>Theorem 5</u>

A quadrilateral formed by the lines joining the middle points of the sides of a rectangles a rhombus whose area is half that of the rectangle.

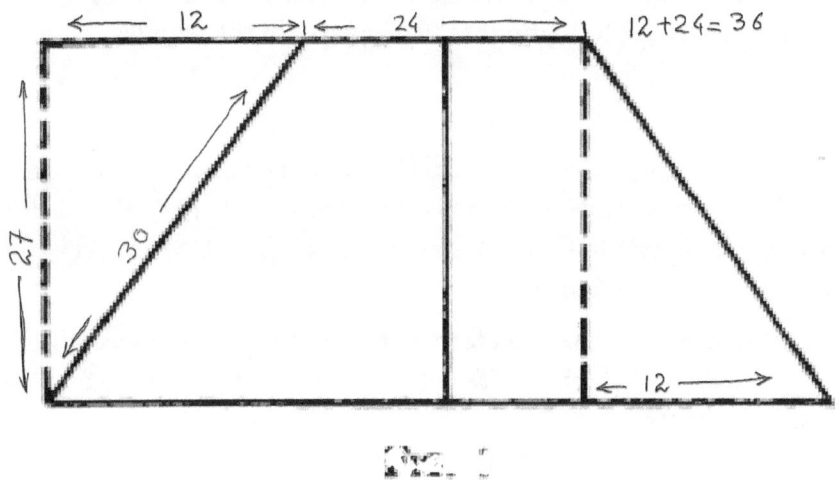

Fig. 3

TEMPLES OF BENGAL

"Geometry of Hindu Temples" will be incomplete
without a separate chapter on "Temples of Bengal". As
a resident of the state where I was born it is my proud
privilege to highlight the geometry of the temples.

I would expect my readers to be aware of the facts that "BENGAL"
the Presidency was divided into two on August 14-15, 1947: East Bengal
which was originally East Pakistan and now "BANGLADESH" after
1971 and the West Bengal.

Although I had the good fortune to visit Bangladesh twice, I'll not
elaborate the "Dhakeshwari Temple" of Dhaka which I visited for the
first time in 1981.

In Kolkata the capital of West Bengal the "Kalighat Temple" is very
famous and the "Dakhineswar Kali Temple" which was constructed
under the auspices of Rani Rashmoni.

As you see the Kalighat temple you will notice a four-sided roof
is being topped with another four-sided smaller roof. Thus total
"Eightfaced-roof" or in Bengali "Aat Chala" temples were constructed
in Bengal alone.

This deviation from the Vastu Geometry is just like the "oniontopped
churches" are the signature of German architecture.

In the district of BISHNUPUR in West Bengal people will see the
one and only "RASHMANCHA" which is topped with pyramid like
dome and the base is rectangular as seen in the picture.

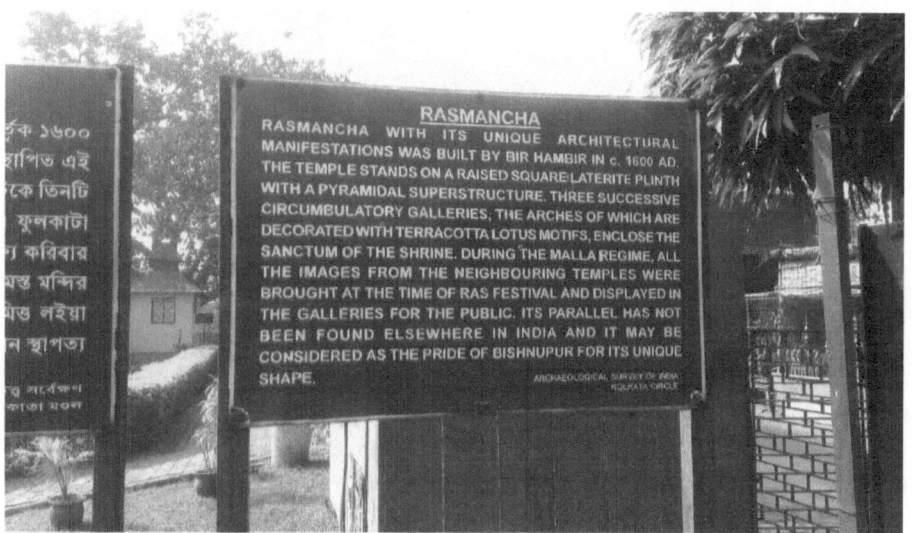

The most recent temple of Bengal is in MAYAPUR, NADIYA DISTRICT, PASHIMBONGO, INDIA.

The temple is called the SRI MAYAPUR CHANDRODOY MANDIR built by International Society for Krishna Consciousness [ISKCON]. The use of marble and the architecture is noteworthy. VRINDAVAN CHANDRODAYA MANDIR IS A TEMPLE UNDER CONSTRUCTION AT VRINDAVAN, MATHURA, INDIA. AS PLANNED, IT WILL BE THE TALLEST RELIGIOUS MONUMENT IN THE WORLD. THIS TEMPLE WILL BE ABOUT 700 FEET [210 METERS] TALL.

Gangaikonda Cholapuram - A grand structure which transports you into our glorious past....!

Built over a 1,000 years back, the gigantic 180-feet-tall Gangaikonda Choleeswaram Mandir is a magnificent testament to the architectural prowess of the Cholas.

Gangaikonda Cholapuram Temple in Gangaikonda Cholapuram, Tamil Nadu

A Chain link hewn out of stone
Sri Varadharaja Perumal Temple, India. Completed 3rd BC.

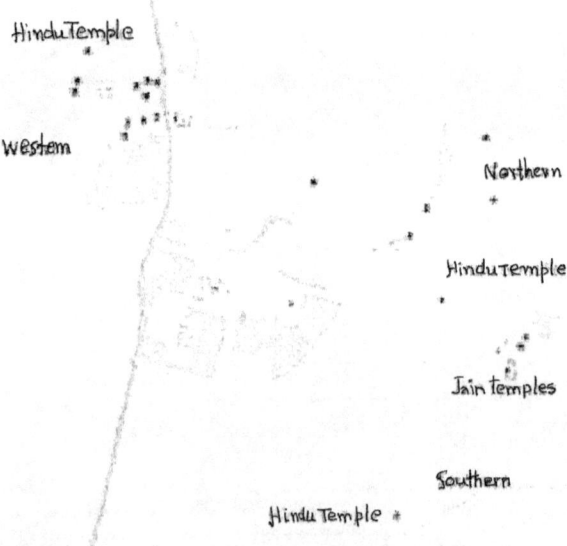

Temples layout map – Khajuraho Group of Monuments.

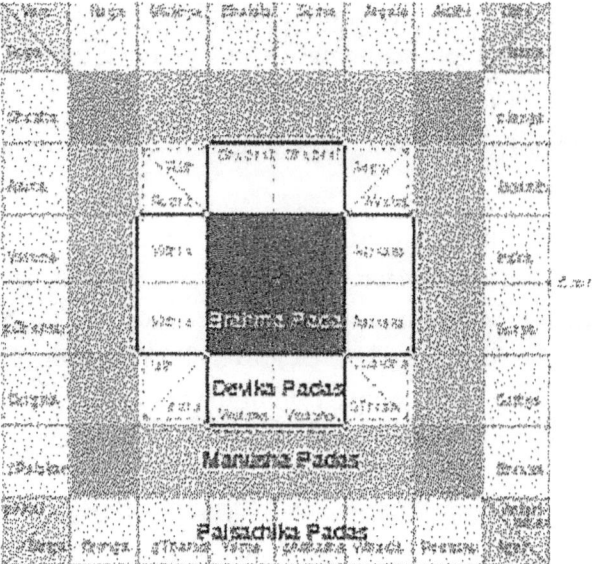

Manduka Mandala - Hindu Temple 64 padas

MATHEMATICISM

Mathematicism is the effort to employ the formal structure and rigorous methods of mathematics as a model for the conduct of philosophy.

MATHEMATICISM is manifested in Western philosophy in at least three ways:

1. General mathematical methods of investigation can be used to establish consistency of meaning and completeness of analysis. This is the revolutionary approach introduced in the first half of the17th century by Rene Descartes. The perfection of this approach led to the age of ANALYSIS in the first half of the 20th century.

2. Descartes also pioneered the subjection of metaphysical systems, expressing the nature of ultimate reality, to axiomatization – i.e., to a procedure that deduces tenets from a set of basic axioms, on the model of Euclid's axiomatization of GEOMETRY. The method was elaborately used later in the 17th century by Benedict de SPINOZA, a Dutch Jewish Rationalist.

3. CALCULI, or syntactic systems, on the model of mathematical logic, have been developed by several 20th century Analytic philosophers, among them Bertrand Russel [ENGLISH], Ludwig Wittgenstein [AUSTRIAN] and Rudolf Carnap [GERMAN], to represent and to explicate philosophical systems, as well as to solve and to dissolve metaphysical problems.

Descartes gave four rules of method in philosophy based on mathematical procedures:

1. Accept as true only indubitable (SELF-EVIDENT) propositions.
2. Divide problem into parts.

3. Work in order from simple to complex.
4. Make enumerations and reviews complete and general.

When a philosopher approaches metaphysical problems in this way, it may appear to be natural or useful for him to organize his/her philosophical knowledge in the form of definitions, axioms, rules and deduced theorems. In this way one can assure consistency of meaning, correctness of inference and a systematic way to discover and to exhibit relationships.

Notes

Notes

REFERENCES

1. Datta, B: Journal of Asiatic Society of Bengal (New Series) Vol. XXVI [1930] pp.283-299.

2. Datta, B and Singh, A.N.:HINDU GEOMETRY, REVISED BY Kripa Shankar SHUKLA, Department of Mathematics and Astronomy, Lucknow University, Lucknow, INDIA [1979].

3. Deva, K : Temples of Khajuraho, 2 Volumes, Archaelogical Survey of India, New Delhi.

4. Desai, D : Khajuraho, Oxford University Press, [2005], ISBN: 978-0-19- 565643-5.

5. Kramrisch, S : The Hindu Temple, Vol.1, Motilal Banarsidass, ISBN:978-81- 208-0222-3, INDIA.

6. Panda, R and Almohammadi, A. R : OUR COLORFUL WORLD IN AJANTA & ELLORA, Mittal Publications, ISBN:978-93-80936-09-3, INDIA.

About the Author

- Born in Kolkata (Calcutta), India. Graduated from University of Calcutta, India with Honors in Mathematics (1974), M.Sc. in Applied Mathematics (1976), M.Phil in Physical Sciences (1981) and Ph.D in Applied Mathematics (1984).
- Ph.D in Statistics (1992), La Trobe University, Melbourne, AUSTRALIA
- Served in the West Bengal Educational Service (Govt. of West Bengal) as Lecturer / Reader of Mathematics.
- Appointed as a Lecturer in Quantitative Methods in Victoria Institute of Technology, Lecturer in Statistics and Operations Research in Royal Melbourne Institute of Technology (RMIT) and Lecturer in Econometrics in Monash University, AUSTRALIA.
- Appointed as adjunct Lecturer/Professor in New Jersey Institute of Technology (NJIT), City University of New York (CUNY) and Kean University
- Delivered Hybrid Internet Course in Statistics for ITT Technical Institute, USA us
- Worked as a Sr. Analyst / Programmer in Paine-Webber, Inc.,QA Manager, iXL, Inc.,Sr. QA Engineer in iNautics Technologies, CSFB Direct, Sr. QA Analyst in e-STEEL Corporation, QA Manager, Navisite Inc., Sr. Consultant, Hexaware Technologies, QA Manager, Blair Corp., SQA Manager, ACNIelson, Senior Consultant at UBS with KEAN, Sr. Specialist, DOD, US Army with Serco and lastly QA Metrics-Analyst with IBM-Walmart.
- Founder FEASIBLE SOLUTION LLC in 1998.